IMAGES
of America

NAVAL AIR STATION
JACKSONVILLE

The first Navy aircraft assigned to Naval Air Station Jacksonville was this Grumman J2F-3 Amphibian Duck. The aircraft arrived at Jacksonville Municipal Airport on January 17, 1940, from San Diego. It was also the first to land on the St. Johns River, near the station, due to a severe storm. Station executive officer Cmdr. V.F. "Jimmy" Grant was at the controls. (Courtesy of the US Navy.)

ON THE COVER: Beechcraft SNB-1 Kansan aircraft are shown on the flight line with officers and crews in formation in March 1943. These aircraft were used for multiengine training at the station. Almost 90 percent of the United States's navigators and bombardiers in World War II were trained in this aircraft. (Courtesy of the US Navy.)

IMAGES
of America

NAVAL AIR STATION
JACKSONVILLE

Ronald M. Williamson and Emily Savoca

ARCADIA
PUBLISHING

Published by Arcadia Publishing
Charleston, South Carolina

Printed in the United States of America

Library of Congress Control Number: 2001012345

For all general information, please contact Arcadia Publishing:
Telephone 843-853-2070
Fax 843-853-0044
E-mail sales@arcadiapublishing.com
For customer service and orders:
Toll-Free 1-888-313-2665

Visit us on the Internet at www.arcadiapublishing.com

*Dedicated to the personnel of NAS Jacksonville; to Joe
Williamson, a great dad and Navy man; and to Dr. Anthony
"Tony" Savoca, a Martin B-26 Marauder bombardier-
navigator with the 320th Bomb Group in World War II*

CONTENTS

Acknowledgments

There are numerous sources, both public and private, that contributed to this publication. In early 1990, there were no actual station "history records" documented at Naval Air Station (NAS) Jacksonville. Today, there is a great array of photographs, historical documentation, and even artifacts. This came about with the assistance of numerous personnel at the station, and especially from former military and civilian employees who were stationed at or worked at the station throughout its almost 75-year history.

First and foremost, I thank former NAS Jacksonville commanding officer Rear Adm. Kevin Delaney. He first supported preserving the history of NAS Jacksonville starting in 1989. Other commanding officers since him who have been very interested in the base history and contributed in some major way include Capt. Charles "Skip" Cramer, Capt. Robert Whitmire, Rear Adm. Steven Turcotte, Rear Adm. Jack Scorby, and Capt. Robert "Colonel" Sanders. The base's public affairs staff of Kaylee Larocque, Clark Pierce, and Miriam Gallet was always available for needed support.

The following individuals helped review selected images, chop text, and provide the necessary guidance for the final product; without their assistance, this publication would have been difficult to accomplish: My deepest gratitude goes out to Jacqueline Kern; Mary Francis Chergi; NAS Jacksonville Command Master Chief Bradley Shepherd; Chris Scorby; Sandra Acosta; Kelley Johnson; Linda Doktor, who scanned all the images; Perry Driver; Marsha Childs; Lt. Cmdr. Max Bassett; Capt. Brett Calkins; Gary Anderson; and my son, Justin Williamson, who is now appreciating Naval Aviation history. Finally, special thanks go to my coauthor, Emily Savoca, for all her contributions.

Unless otherwise noted, all images are provided courtesy of the US Navy.

INTRODUCTION

The Naval Air Station (NAS) Jacksonville site has a long history. Early inhabitants, of what is known as the Early Archaic culture, were nomadic hunters and gatherers who occupied the site in short-term camps as early as 6500 BC. From this period until European contact in 1565, numerous aboriginal populations visited the site. One location on the base has yielded more than 9,500 artifacts.

Three Spanish grants account for most of the early European history. The first grant, Pointa Negra, or "Black Point," was given to Don Felipe Bastros. The second grant, Sans Souci, was granted to A.C. Ferguson, and the third, Moral Grueso, meaning "Fine Mulberries," belonged to T. Hollingsworth. Alfred M. Reed eventually bought most of the property in 1862, and it became known as Mulberry Grove Plantation. The onetime plantation is now occupied by the officers' housing area. Reed, his wife, and two daughters lived in a large home overlooking the St. Johns River. Reed died in 1886, and, in 1905, family members began selling off parcels of Mulberry Grove Plantation, with the last 336 acres of the original 1,400-acre plantation sold for the establishment of NAS Jacksonville in 1939.

The rich military history of NAS Jacksonville began in 1906, when a search for the new site for the state militia began. After investigating many sites in Florida, a 1,300-acre tract of land at Black Point, known then as Philbrofen, was finally recommended in 1907. The first tract of land, of 389 acres, was acquired from Joseph H. Phillips for $8,000. The citizens of Jacksonville raised $6,000 to establish the camp, and the federal government made available $8,000 to purchase additional properties. The first encampment of state troops at Black Point was held on June 8–15, 1909.

The site's aviation history began on December 4, 1916, when New York millionaire Earl Dodge opened an aviation training camp on the site, which today houses the offices of the base commanding officer. Aspiring pilots were initially trained in three Curtiss hydroaeroplanes before moving on to JN Jenny aircraft. Of note, not one single aviation accident occurred during the almost two years of training.

The Army started looking for sites for cantonments—large training camps—in 1916. Gen. Leonard Wood, who was to pick sites for 16 cantonments, sent an aide to inspect the Black Point site. The day that aide arrived was rainy, and he left with a most unfavorable recommendation. However, W.R. Carter, the editor of the newspaper *Jacksonville Metropolis*, decided to fight for the establishment of the camp. His efforts prompted General Wood to make a personal trip on June 25, 1917. He was so impressed that he returned to Washington and Jacksonville was selected. Even this recommendation went back and forth, though, and it was not until Gen. Francis J. Kernan, a West Point graduate and Jacksonville native, endorsed the site that the camp was finally secured.

On September 10, 1917, J.S. Pray, a professor of landscape architecture at Harvard University, was hired by the War Department to lay out the camp. One condition for the selection of Jacksonville was that liquor be kept away from the soldiers stationed there. In September 1917, the War Department took over control of the property, and, one month later, construction on the new Army camp commenced. Camp Joseph E. Johnston, named after a Confederate Civil War general, was commissioned on October 15, 1917, to train soldiers for World War I. A shooting range was added later where the airfield is today. At the time, it was the second-largest rifle range in the United States.

The first soldier arrived at the camp on October 16, 1917. He was actually sent there by mistake from Camp Custer in Battle Creek, Michigan. Construction had just started, so the contractors located the only officer present, Col. Fred Munson, who fixed him up with quarters. The first group of officers and enlisted men arrived for training on November 19, 1917, and, by December 3, 1917, the men had left for the battlefronts in Europe. In December, the camp was selected as a remount station and 160 acres in Yukon was cleared.

The depot consisted of 16 buildings and 14 stables and had provisions for 4,000 horses and mules. The population of the camp peaked at 27,000 men, most of them departing from Black Point after a brief training period for service in Europe. By June 1918, an additional $1.7 million expansion project was submitted that would have built quarters and facilities to train 50,000 men. That project was canceled when World War I ended on November 11, 1918. Within two weeks, the camp was ordered demobilized. By February 1919, only a few soldiers remained to guard the camp. Camp Johnston was officially closed on May 16, 1919. Shortly after that, the base hospital, which sat on 126 acres of land, was transferred to the Treasury Department for public health service.

The state militia tried to get the property back for use almost immediately after the end of World War I. However, the Army continued to hold on to the property. The aviation field, used for training in World War I, was the site of a new transcontinental speed record when Maj. Theodore C. McCauley landed there on April 18, 1919, having flown from San Diego in 25 hours and 45 minutes.

The Army abandoned the camp on July 23, 1920. On June 22, 1921, a fire broke out in the military warehouses in Yukon and destroyed all of the equipment the Army had stored there. Three days later, the federal government gave the state title to 682 acres and 154 buildings. The remaining 458 buildings and supplies were sold at public auction for a fraction of their value. On June 7, 1926, the War Department finally granted a revocable license to the State of Florida so the lands could be used once again for National Guard training. The state took control and, on June 18, 1928, renamed the site Camp J. Clifford R. Foster after the adjutant general of Florida in the early 1900s, who had recently died. His actions in 1905 led to the establishment of military activities at Black Point.

On February 22, 1934, a motorcycle race was held on the station grounds. Today, the race is known as the Daytona 200 Motorcycle Race. Also during this time, the first air show was held on the grounds. In 1935, when the Army Air Corps was looking for a site for an air base, Jacksonville politicians made a pitch to locate the base here. They lost that battle when the Army picked Tampa for what is known today as MacDill Air Force Base.

In May 1938, Congress created a board to determine suitable sites for naval shore facilities. Headed by Rear Adm. A.J. Hepburn, it was known as the Hepburn Board, and it first visited Jacksonville in May 1938. On October 7, 1938, three seaplanes arrived and began landing and takeoff tests in the St. Johns River. This was the first indication the Navy might be serious about using the Jacksonville site. The second visit by the Hepburn Board was on November 8, 1938. On January 9, 1939, House Rule 2880 was introduced into Congress, which included the authorization for NAS Jacksonville. However, other cities competing for the new naval base complained, and another Hepburn Board review was recommended. A third visit was conducted by the board on March 2, 1939, and finally, on March 22, Jacksonville was recommended as the preferred site for the new base. A bond issue to support the purchase of the additional property was approved by the citizens of Duval County by a vote of 13,808 for and only 265 against.

Construction began almost immediately when $15 million was allocated by Congress on April 26, 1939. The Armory board was not in favor of the deal, however, as it meant that the Florida National Guard site Camp Foster had to find a new home. The board finally requested compensation, which it got in the amount of $400,000, to move to a new site in Clay County, the current home of Camp Blanding. On October 23, 1939, construction of Camp Blanding began. So complete was the move from the Jacksonville location that even most of the manhole covers were removed. Condemnation suits started on November 23, 1939, as some property owners were not happy at the thought of losing their land. Title to the property was transferred on April 23, 1940, making NAS Jacksonville the only military installation in the United States gifted directly by the people to the government. However, of the 243 parcels condemned to allow the acquisition of the lands, 41 were still to be settled as of July 1940.

At high noon on October 15, 1940, NAS Jacksonville was commissioned as a primary flight-training base, with Capt. Charles P. Mason selected as the first commanding officer and with Adm. John Towers in attendance. N2S Stearman biplanes arrived in December, and flight training commenced on January 2, 1941. Training in PBY Catalina seaplanes began in June 1941. The station grew quickly. By the end of 1941, almost 600 aircraft were flying from the runways. Additionally, a base at Green Cove Springs (now Lee Field), a naval auxiliary air station at Cecil Field, and facilities at Mayport were also constructed and placed under the control of NAS Jacksonville.

Numerous other outlying fields and bombing ranges were also scattered across northeastern Florida for use by the station's aircraft. Walt Disney designed NAS Jacksonville's first three official logos, with the one of Donald Duck's nephews coming out of an eggshell and learning how to fly being the most popular. Although the use of that logo ended in 1943, it was readopted by the station in 2000 and can be seen today on the station's fuel tank on Yorktown Avenue as well as on the caps worn by the station sailors. Some 4,363 fledging airmen came through NAS Jacksonville and earned their Wings of Gold as pilots before primary flight training was transferred to the newly established base in Corpus Christi, Texas, in late 1942. The last graduates to be awarded their wings graduated on February 26, 1943. The station then became an advanced fighter-pilot training base until the end of World War II. Additionally, the Naval Air Gunners School at Yellow Water also trained more than 30,000 gunners from 1941 through 1945.

The Assembly and Repair Department was also formed with the creation of the station. This department initially assembled aircraft that arrived, usually by train. Those in this department also repaired all of the station's aircraft. During the war, almost 1,000 aircraft were repaired and overhauled in the facility.

As home to the Naval Air Advanced Training Command, and with so many fighter pilots based here after World War II, NAS Jacksonville was the logical choice to establish a new flight demonstration team, known today as the Blue Angels. Capt. Roy M. "Butch" Voris was selected to form and lead the team. Its first show in front of Navy officials and base personnel was flown on June 7, 1946, followed shortly by its first public demonstration, during the dedication of Craig Field in Jacksonville on June 15, 1946.

The second half of the 1940s saw many military bases close, but not NAS Jacksonville, which remained viable because it was home to Carrier Air Group's basing fighter and attack aircraft. In late 1949, Fleet Air Wing 11 (now Patrol and Reconnaissance Wing 11) and the first three patrol squadrons started their transfer to the station, with Patrol Squadron VP-5 being the first to relocate. The VP-5 "Mad Foxes" arrived in December 1949 and remains the oldest squadron continually based at the station.

The 1950s saw continued change and the introduction of jet aircraft. The Overhaul and Repair Department, formerly known as the Assembly and Repair Department, was busy reworking Navy helicopters and the S2F Tracker and inducting many new jet aircraft.

The airfield was named Adm. John Towers Field during the station's 20th anniversary, on October 15, 1960. This made NAS Jacksonville unique in that the station name and the airfield names were different. The Overhaul and Repair Department became the Naval Air Rework Facility, a separate tenant command, on April 1, 1967.

Base growth continued in the 1970s when Helicopter Antisubmarine Wing, US Atlantic Fleet, and its assigned helicopter squadrons transferred to the station. That transfer brought seven new helicopter squadrons to the station: HS-1, HS-3, HS-5, HS-7, HS-11, HS-15, and HC-2. Eventually,

HS-9, HS-17, and HS-75 were also added. On March 8, 1974, VA-203 would retire the Navy's last A-4 Skyhawk at NAS Jacksonville. April 30, 1975 saw the disestablishment of the Navy's Hurricane Hunters.

With the announced closure of NAS Cecil Field in 1997, Sea Control Wings Atlantic and its six S-3 Viking squadrons started the transfer to the station. Soon to arrive would be VS-22, VS-24, VS-30, VS-31, VS-32, and VQ-6. The air station went on to acquire Outlying Field (OLF) Whitehouse and the Pinecastle, Rodman, and Lake George bombing ranges, with the final closure of NAS Cecil Field on September 30, 1999.

The Viking squadrons did not have a long history at NAS Jacksonville. By January 29, 2009, with the disestablishment of VS-22, all were gone. Although the VS squadrons were being disestablished, five new patrol squadrons from the closing NAS Brunswick were arriving at the station.

On March 28, 2012, the formal arrival ceremony and induction of the P-8A Poseidon patrol plane was conducted. This aircraft will eventually replace the P-3 Orion aircraft currently operating at the station. VP-16 began training in the P-8A as the first operational squadron on July 11, 2012. It is projected to take almost five years for all of the P-3 Orion aircraft to be phased out of operations at NAS Jacksonville. The day before the P-8A arrived, the MQ-88 Fire Scout operator-training facility was unveiled. This unmanned aerial vehicle will be used in conjunction with the helicopter squadrons at the station.

Today, NAS Jacksonville is at the forefront of the global war on terror. The current SH-60 helicopter squadrons are being redesignated as Helicopter Squadron Combat (HSC) and moving to Naval Station Norfolk and NAS North Island. New Helicopter Maritime Strike Squadrons (HSMs) are taking their place. This master air and industrial base will be home to 16 Navy operational, training, and reserve squadrons by early 2015, the station's 75th anniversary.

One

CAMP JOHNSTON AND CAMP FOSTER

The first officers assigned to Camp Joseph A. Johnston are seen here shortly after their arrival on November 2, 1917. Maj. Fred I. Wheeler, the construction quartermaster (seated, left) was in charge of camp construction. Major Wheeler stayed until construction was completed on April 22, 1918. He and two other members were known as the Three Musketeers around the camp, and his staff adopted the saying "One for all and all for one" as they oversaw camp construction.

After the clearing of sections of the camp of all trees and shrubs, temporary tents were initially set up for Army personnel arriving while construction was taking place. Problems with drainage, heat, and mosquitoes meant that these initial living conditions were not the most pleasant for the new arrivals.

This narrow gauge railway was built from the main line, running along what is now Yukon, so construction materials could be moved to build the new $2.9 million Camp Johnston. Approximately 29,800 feet of temporary narrow gauge industrial track was laid throughout the camp. The workers that built the camp were employees of A. Bentley & Sons Construction of Toledo, Ohio.

Construction was in full swing on October 26, 1917, as seen in this photograph showing the construction of a building's floor system. All buildings had to be raised due to the swampy conditions created during periods of heavy rain. Local Jacksonville workers were hired and transported by ferry every morning.

Once buildings were raised to get above the swampy grounds, workers rapidly built the walls and frame roofs. Structures such as this one were usually completed in less than a week. Over one million feet of lumber was used to construct the camp buildings, some of it coming from the trees felled while clearing land for the camp.

Tons of fill dirt was hauled in to raise roads. Almost all of the buildings, like this barracks, were placed on stilts due to the swampy site conditions. Additionally, 194 bridges were constructed across the camp's low areas. The first power lines, seen alongside the roadway, were also run to the buildings to provide electricity for lighting.

2604 CLUB HOUSE OR HOSTESS HOUSE. CAMP JOHNSTON. FLORIDA.

The Hostess House at Camp Johnston, located on the banks of the St. Johns River, was used by camp officers. On Sunday afternoons, the officers could invite local Jacksonville women to have an afternoon dinner. Officers would dress in their formal attire, as it provided a degree of sophistication to the camp life.

2601 BACK OF HOSTESS HOUSE ON ST. JOHN'S RIVER, CAMP JOHNSTON, FLORIDA.

The Hostess House, also known as the dance pavilion, provided a porch where officers could relax and enjoy the view and the afternoon breeze off the St. Johns River. Demolished in the early 1920s, it was located near the end of the seawall between what are now Hangars 124 and 140.

Two theaters were established on the camp grounds for the entertainment of the troops. Built in less than a week and nearly identical, both the American and Liberty Theatres were very popular on Friday and Saturday evenings. Both theaters showed the latest movies and occasionally had live entertainment for the camp troops.

15

A well-equipped library was also located at the camp, where soldiers could read the latest newspapers of the day, including the *Jacksonville Metropolis*. Seen here behind the library is one of five Knights of Columbus halls that also provided entertainment and a place for socializing for the camp soldiers. The Knights of Columbus developed into a fraternal service organization dedicated to providing charitable services and promoting Catholic education.

Camp Johnston was established as a quartermaster camp. The area west of Route 17 (Roosevelt Boulevard) and behind Yukon was where the horse barns were established for horses and mules. A cart barn is seen here. Almost 4,000 horses and mules were kept in this section of the camp.

Seen here is Fire Station No. 2, located in the middle of the camp. There were three fire stations at the camp, all equipped with the latest firefighting equipment. There were no major fires reported during the time training was conducted at the camp. Firefighting training was important because all of the structures were made of wood and the new electrical installations did not have today's safety codes. The chance of a fire was considered high. The letters on the front of this fire truck stand for "Quartermaster Camp United States Army." Brick roads like this one were located throughout the camp and at the Yukon site. Recent efforts have been made to uncover some of the Yukon brick roads, which are buried under forest debris today.

The above panoramic view shows the main section of Camp Johnston in March 1918. When the camp was completed, there were 549 buildings, 48,016 feet of water lines, 47,269 feet of sewer lines, 3 miles of railroad track, 17 miles of pole lines, 5 miles of brick streets, and 6 miles of fences. The

photograph below, taken on March 4, 1918, shows the approximately 3,000 men at the camp in formation. The camp band is on the far right.

Camp Johnston had an Army band that played at events and ceremonies held on the base. The band also gave very popular Sunday afternoon performances at the parade grounds. The band was also invited numerous times to play at events in downtown Jacksonville.

The main water tower at Camp Johnston was constructed in approximately the same location as the base's current water towers. This tower supplied all of the water needs of the camp. The wooden construction shows the massive timbers used to support the weight. This particular water tower stood through the 1930s before being demolished to make way for a new one.

The camp's rifle range was the second-largest in the United States. Completed in early 1914 by the state militia, it had ranges for rifles as well as machine guns. There were 157 targets on the range. National rifle competitions were held here in October 1915. The Army took over the range when it used the property. The range is under the current site of the NAS Jacksonville runways.

Boxing matches were held on the camp grounds for the entertainment of the soldiers. Here, Jess Willard (left) and Jack Phelan square off for the assembled soldiers to watch. Willard, the world heavyweight champion at the time of the fight, easily defeated Phelan.

BANDSTAND. CAMP JOHNSTON. FLA.

The bandstand was located near the parade grounds. The camp band played at the bandstand, which also provided a viewing area for the parade grounds. Many of the trees cut to make the camp were used in the construction of this and many other facilities on the grounds.

ST. JOHN'S RIVER IN FRONT OF CAMP, THREE MILES WIDE HERE, CAMP JOHNSTON. FLA.

The pier, over 257 feet in length, was located on Black Point, where the Kemen Engine Test Cell is today. Steamboats would routinely pull up to the pier. During camp construction, men were ferried in and departed daily from this pier.

The main gate of Camp Foster was about a half mile east of what is now the NAS Jacksonville main gate. The camp at that time incorporated about half of the property NAS Jacksonville occupies today. The two cannons on the roofs of the buildings were German World War I war trophies brought back from the battlefields in Europe. The cannons were relocated to Camp Blanding, Florida, on January 17, 1940, and the entrance was demolished shortly thereafter.

The main flagpole at Camp Foster was east of where the station marina is today. This postcard shows the ceremonial flag being lowered in the evening, with saluting men next to the cannons in the background.

CAMP J. CLIFFORD R. FOSTER
FLORIDA RIFLE RANGE

The main entrance into Camp Foster is seen here, looking east to the St. Johns River. According to the September 9, 1929, inventory, there were 93 structures at the camp and the total property was valued at $835, 318.

Two

NAVAL AIR STATION JACKSONVILLE CONSTRUCTION AND WORLD WAR II

This aerial view from December 1942 shows the main group of Supply Department buildings at the station. They are, from the upper left to the lower right, the commissary and cold storage plant; Building 111, annex with engine and propeller storage; Building 110, main supply building; Building 109, major structural spares storage hangar; Building 160, tire storage; Building 137, lumber storage; Building 108, oil, paint, and compressed gas storage; and, in the foreground, the station administration building.

Construction started at the base in October 1939. This photograph shows the ground floor being laid for Building 110, the three-story supply building. In the center background is Building 27, which remains today. It is the only structure remaining from the Camp Johnston days. A caretaker's cottage is at the top left. This structure was demolished a month after this photograph was taken. The base commander's building is located there today.

The seaplane hangar's steel framework is seen here in October 1940. Three identical hangars were constructed—Hangars 122, 123, and 124. The hangar in the background, Hangar 124, is the only original seaplane hangar still remaining at the station today. PBY seaplanes were originally kept in hangar 124 when maintenance was needed.

This building, located near where the commercial gate is today, was once a country store and a filling station for the town of Yukon. When the Navy took over the compound, it became the first Chiefs' Club on base after some redecorating. It was demolished by mid-1943, and the club moved to a new location.

The station was formally commissioned on October 15, 1940, at high noon. Seen here, from left to right, are Rear Adm. John Towers, Capt. Charles P. Mason, and Cmdr. Carl H. Cotter. Building 110, the first completed building at the station, opened a day later. Captain Mason was the station's first commanding officer and had a long career with the Navy. After he retired, he was elected mayor of Pensacola.

Seen here touring the flight line at the station on March 20, 1941, are, from left to right, Pres. Franklin D. Roosevelt; Capt. Charles P. Mason, the NAS Jacksonville commanding officer; and Rear Adm. Ross T. McIntire, the White House physician and surgeon general of the Navy. President Roosevelt departed the station by train and headed to the Gulf Coast for a fishing trip.

Trade school students learn how to sew rib stitching on the wings of the aircraft in the fabric shop at NAS Jacksonville on November 26, 1941. Fabric stitching was an important skill, as many early aircraft had fabric wings.

Aviation training was the main purpose for the establishment of NAS Jacksonville. Training activities for young cadets started in the Stearman aircraft and then moved on to the NR-1 Ryan. The station's 100 NR-1 aircraft all arrived in July 1941. They remained at the station until the spring of 1943.

The odd-shaped Building 141, seen here in June 1943, was the celestial navigation training facility. It was located north of Hangar 140 and east of where the Fleet Readiness Center Southeast engine facility sits today. All aviation cadets went through this facility to learn navigation skills and night-navigation techniques. Star patterns could be displayed in the four tall areas of the facility. Later converted into a warehouse, it was eventually demolished in December 1969.

These two photographs of construction progress show the western and middle sections of the base on February 15, 1941. The photograph above shows what was designated as the trade school area, where most of the classroom training took place, as well as the new housing area (top). The photograph below shows the ground school area.

This construction photograph, also taken on February 15, 1941, shows the eastern section of the base, also known as the industrial area. Some of the original roads from Camp Johnston can also be seen traversing the middle part of the photograph. Most of the trees were cleared from the base for construction. The majority of the wooden structures in these photographs have been demolished today, including two of the three seaplane hangars, shown at the bottom of the photograph.

This aerial view of Naval Hospital Jacksonville was taken on February 10, 1942. The hospital consisted of a series of one-story wood-frame buildings at that time. With some additions through the years, this remained as the main base hospital until the present hospital was opened in 1967. All of the structures in this photograph were then demolished, with the final structure torn down on July 24, 1968. The four houses at the bottom of the photograph, along the banks of the St. Johns River, were quarters for the senior medical officers of the hospital. The main hospital administration building is seen at the front of the complex.

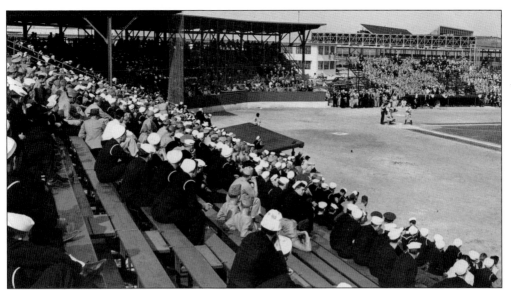

NAS Jacksonville had one of the best baseball stadiums in Jacksonville during World War II. The first game played at the stadium was against the University of Florida Gators. The station team went on to beat the Gators twice in 1942. This photograph shows a game against the Boston Braves on April 7, 1942. The field was dedicated as Mason Field on May 27, 1942, in honor of the station's first commanding officer, Capt. Charles Mason, a baseball fan.

Three of the station's top baseball players in 1942 are, from left to right, Walter Shinn, a former All American lineman at the University of Pennsylvania who batted .342; Pat Calgan, a minor-league catcher with the Louisville Colonels who batted .383; and Bob Leahy, a former Boston College athlete who batted .320. Lt. George Earnshaw, a former Philadelphia Athletics star, was the team's coach. When this photograph was taken on July 24, 1942, the team had 38 wins, 1 tie, and 12 losses.

Seen here are members of the NAS Jacksonville Flyers football team. Base commanding officer Capt. Charles Mason is in the center of the third row. The station did not have a team in 1942 because of the war effort. One was reformed in 1943 and there were teams continually until 1952.

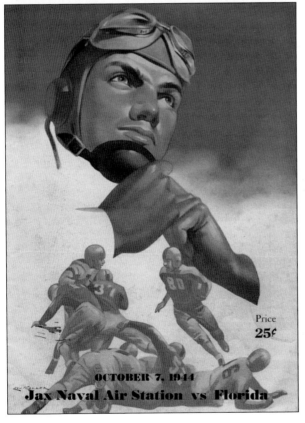

Price
25¢

OCTOBER 7, 1944
Jax Naval Air Station vs Florida

An official NAS Jacksonville Flyers Navy football program, costing 25¢, is seen here. On October 7, 1944, the Flyers played the University of Florida Gators in Gainesville in their season-opening game, losing 27-20.

Members of the Inter-American Defense Board, composed of representatives of 17 South American republics, crane their necks on the NAS Jacksonville seawall to watch 254 station aircraft pass by in review on May 24, 1942. This was the largest aircraft armada ever seen in the skies over Jacksonville and involved all the station's N2S Stearman, NR-1 Ryan, and SNJ and PBY Catalina aircraft.

Part of the physical readiness training provided for all naval personnel at the station involved learning how to use a rifle with a bayonet. Below, Maj. Dick Hanly (center left) and Lt. Col. William T. Evans of the Marine Corps (center right) watch as a student demonstrates his skills.

Flight cadets are seen here learning trap shooting on September 10, 1942. The same technique is needed in aiming and following with a machine gun mounted in a fighting plane. Cadets practice on the range for many hours before using guns mounted in planes. The man firing has just scored a perfect hit, as evidenced by the black puff of smoke. He stands in perfect form as the cartridge pops from his gun.

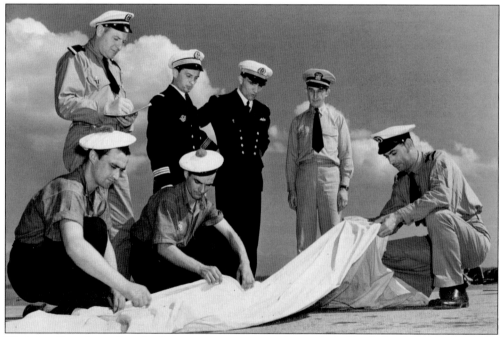

"Fighting Frenchmen" at the base examine a target sleeve for bullet holes following gunnery practice as part of their training at the station on November 2, 1942. French sailors were easily recognizable by the red topknots on their hats. A base naval officer stands in the background.

Naval officers relax in the lounge at the senior officers' quarters, Building 11, on Mustin Road on the base, on November 2, 1942. The historic mural along the top of the wall depicts the history of Florida. Building 11 is currently scheduled for future demolition. In 2012, efforts to remove and relocate the mural were undertaken. Today, the mural resides in the Museum of Science and History in downtown Jacksonville.

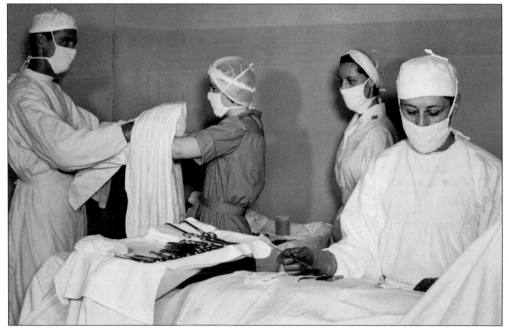

Women Appointed for Voluntary Emergency Service (WAVES) are seen here at Naval Hospital Jacksonville on July 25, 1943. Hospital apprentice first class Hazel Brown (second from left), under the supervision of Ens. Margaret Mitchell (third from left), is being initiated into operating-room techniques by preparing to assist in an appendectomy. The WAVES performed many functions at the naval hospital during World War II.

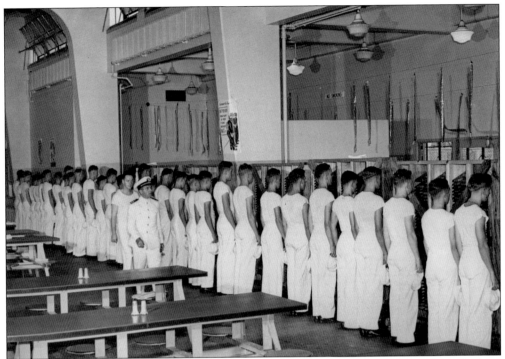

The permanent officer-of-the-day inspects mess cooks prior to them serving meals in the base galley in July 1943. After inspection, the cooks prepared soup, fry steaks, and baked potatoes for noon chow. Thousands of military personnel ate at the galley daily.

Military pay lines start to form outside Building 954 on regular paydays, where there are three such lines. When this photograph was taken in July 1943, regular payday was of three days duration. Today, military pay is issued twice a month. The two-story wooden building in the background is of the typical "H" design constructed throughout Navy bases during World War II.

Building 160, seen here in July 1943, housed the major salvage stores section. This general view down the east one-third of the building shows damaged aircraft parts. The center section shows damaged aircraft propellers. All of the materials brought here are salvaged for reuse in overhaul and repairing aircraft assigned to the station and to the outlying stations of Lee Field, Cecil Field, and Mayport.

Station aircraft used a lot of fuel and lubricants. Fuels were brought in by barge. Because of the proximity of the aviation lubricating oil storage area to the fuel storage area, Navy Symbol 1100 and 1200 oils are handled by fuel division personnel rather than oil and paint section personnel. Here, an inspection department inspector and a fuel division storekeeper inspect an arriving tank car of aviation lubricating oil.

Four Naval Academy graduates of the class of 1943 watch intently as an instructor explains a fire-area projection device. Suspended in the center of a hollow sphere is a model of a Japanese plane, indicating the angles and areas from which a United States plane could attack without encountering enemy fire.

Lt. Comdr. Ronald Higgins, the commanding officer of the Aviation Service Schools, and Capt. Charles Mason, the commanding officer of NAS Jacksonville, conduct a review of the naval personnel assigned at the station schools on April 25, 1942. A captain's flag presentation was also conducted for the class of graduating personnel and presented in front of the invited guests and family members of the cadets.

Three WAVES aviation machinist mates work on an SNJ training plane on the station ramp on November 2, 1943. All were graduates of the Naval Air Technical Training Center in Norman, Oklahoma, where they spent four months learning to repair and overhaul aircraft engines. Most of the station WAVES machinist mates were assigned to the Assembly and Repair Department.

Students using bucket-seat machine guns take practice on November 17, 1943. This training took place at the Naval Air Gunners School, located at Yellow Water, across from Naval Auxiliary Air Station Cecil Field. This was just one of many different types of guns used for training by the students at Yellow Water. Some 30,000 gunners were eventually trained. These students were designated aviation ordnancemen "T"—for turrents.

The aircraft depth bomb Mark 54 is being transported for loading on a PBY Catalina. Like its close relative the depth charge, this depth bomb is unlikely to hit a submarine directly. Instead, it creates an underwater pressure wave in an attempt to weaken or crush the hull plates of the target. The practice of riding the bombs as seen here is not allowed today under the Navy's strong Explosives Safety Program.

Once the depth bombs are winched into place, the work platform is removed by two enlisted personnel as a Navy chief supervises. With 100 PBYs eventually assigned to the station, the weapon's department personnel were very busy loading and unloading these depth bombs on Catalina aircraft. Not one incident involving these weapons occurred at the station during World War II.

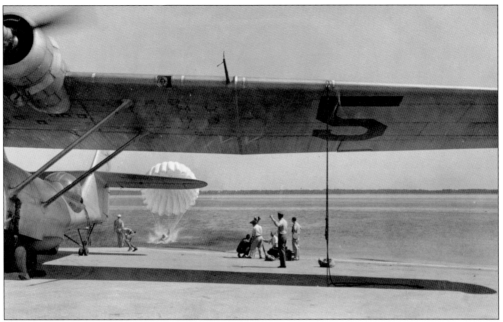

This photograph shows the filming of a movie using a Consolidated PBY Catalina in June 1943. Numerous training films were shot at NAS Jacksonville during World War II. Here, the effects of propeller wash on a parachute are being simulated. Movies filmed at the station were then distributed throughout the Navy as training aids.

The last class of aviation machinist's mates is seen here in December 1943. The mates maintained the engines, gearboxes, rotor blades, and propellers of aircraft. They could also manufacture any specialty pieces that were needed. This school was moved from NAS Jacksonville to NAS Memphis on January 8, 1944.

Planes involved in crashes and awaiting major overhaul by the Assembly and Repair Department or that have salvageable parts, equipment, or structures awaiting removal are brought to the salvage plane storage corral. Obsolete aircraft are also brought to the corral, pending salvage operations. This is a general view of the storage corral in January 1944.

Supply personnel remove a preserved engine from an FM Wildcat at the storage corral. Once removed, it would be placed in the engine storage section until the Assembly and Repair Department could overhaul it. After being overhauled, it would be ready for installation on an overhauled aircraft or on one in need of an engine.

Sometimes the market for scrap was so glutted that bids were too low to make salvage operations profitable. In such times, aircraft, like these FM Wildcats, were placed in a vacant field where all usable parts and equipment were removed. The aircraft were steam-cleaned of all oils and lubricants and the wings were removed. Then they were trucked to the base pier at the end of Albemarle Street, where they were unloaded by a crane and transferred to a barge taking them off the coast of Jacksonville to be dumped at sea. This practice was eventually stopped when the citizens of Jacksonville complained.

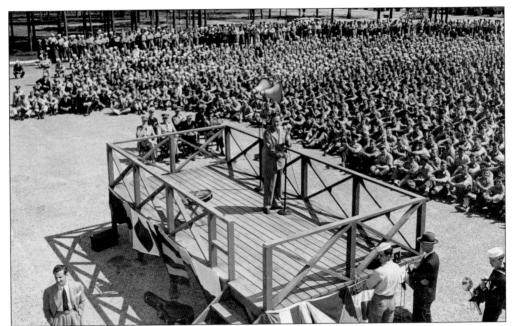

A coast-to-coast radio broadcast was transmitted from the station on March 14, 1944. Bob Hope performed prior to the winging ceremony in front of personnel at the Naval Air Technical Training Center. During the ceremony, 150 airmen received their Wings of Gold.

Bob Hope converses with Rear Adm. Andrew C. McFall at the NAS Jacksonville Officers' Club following Hope's performance in March 1944. Rear Admiral McFall was in charge of the Naval Air Operational Training Command (NAOTC) and was the senior officer at the station, even over the station commanding officer. NAOTC was responsible for all air training in the southeastern United States at the time.

The Naval Air Technical Training Center post office was a busy place around Christmas, as seen in this photograph from December 20, 1943. On the far left sorting mail is Gail W. Alrius, and to her right is Marjorie Schuly. Mail was important to the personnel assigned to the station, and military personnel could send mail for free.

The WAVES service shop was located in the base's shopping complex. Here, WAVES could buy personal items as well as uniform articles. This photograph was taken on March 20, 1944.

Physical training was a big part of the daily lives of all students at the Naval Air Technical Training Command. Here, on March 21, 1944, students are taught how to inflate their trousers to make life vests. This technique is still taught today.

In addition to general fitness training, students were taught boxing skills, as seen in this photograph of instructors leading a large class on May 1, 1944. Fighting skills were also taught in other areas involving knives and guns with bayonets.

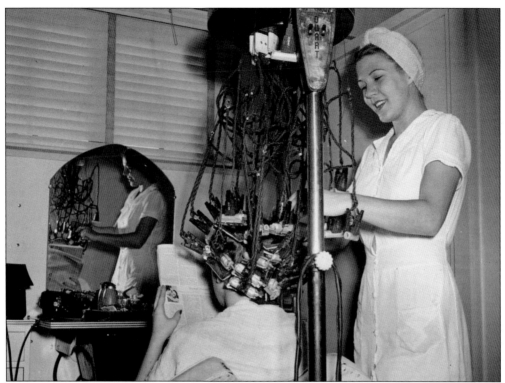

The WAVES beauty shop, seen here on November 2, 1944, was the place on base to have one's hair done. It is unknown if this machine would even pass the safety standards of today.

WAVES personnel are being taught CPR skills by the base pool in this 1944 photograph. Basic lifesaving skills were also taught to all of the WAVES as part of their general military training.

This bond rally took place at the Naval Air Technical Training Center on April 26, 1944. Raising money for bonds was a huge campaign at the station during World War II. Personnel always seemed to surpass any monetary goal set.

These carts were used for student training at the bombsight school. As seen here on May 24, 1944, students drove the carts to the target and then hit a button releasing a sandbag that dropped onto the target. Today, the hangar, now belonging to the Morale, Welfare, and Recreation Department, is used by sailors working on their vehicles as the station auto hobby shop.

Rear Adm. Andrew C. McFall, commander of the Naval Air Operational Training Command, approaches the band shell for a presentation of the war bond pennant to the commanding officer of NAS Jacksonville, Capt. Charles Mason, on May 29, 1944. The band shell, located just west of Building 1 on the base, was used for many events until it was finally demolished in November 1966.

Senior officers of the Naval Air Operational Training Command are seen here on January 16, 1945. This group of officers commanded 16 major naval air stations, 4 aircraft carriers, and more than 65,000 naval personnel. The Naval Air Operational Training Command, located at the station until November 1, 1945, was later renamed the Naval Air Advanced Training Command.

This PBY Catalina was stripped and used as an outdoor classroom for ordnance personnel. Known as the Gravel Gertie, it was used to train ordnancemen how to load and unload the 500-pound depth bombs under the wings of a PBY. During the time the Gravel Gertie was in use, hundreds of naval ordnancemen learned the art of safely handling and loading depth bombs. The aircraft was finally scrapped in late 1946.

This hulk of a PBY Catalina was used for ditching drills. The aircraft was launched off of the base seawall and floated out into the St. Johns River, as seen here on January 18, 1945. Once it was anchored in place, boats would transport aviators and aircrew members and place them inside the aircraft. From there, they could practice plane-ditching drills by escaping out of the plane and jumping into the St. Johns River.

There were many schools teaching a variety of skills at the station during World War II. One of the smaller but more important ones for aviators was the aircraft tire-retreading school. Here, a sailor works on a tire for a PBY Catalina patrol bomber on November 24, 1944.

The first ingot from the new aluminum-melting furnace is held by Ens. Robert Brigleb, station conservation officer, as station salvage officers watch on January 5, 1945. Thousands of aircraft frames, pieces, and parts were melted down after the war, and the aluminum ingots were then sold to private bidders.

The city of Jacksonville's military train reservation booth was located in the downtown terminal. Most Navy personnel that arrived for training at the station came through this terminal. Another terminal stop point was actually in Yukon, across from the station's main entrance. But the main terminal in downtown Jacksonville was where the vast majority of new recruits entered Jacksonville. This photograph is dated January 13, 1945.

Students get instructions on assembling, cleaning, and disassembling a .50-caliber machine gun at the Naval Air Technical Training Center on July 12, 1944. This gun was one of the dominant types used by the fighter aircraft attached to the station. Aviation ordnanceman "A" and "C" schools were taught at NAS Jacksonville and were 20 weeks long. Today, aviation ordnanceman schools are located at NAS Pensacola.

Students assigned to the physical training department receive instructions for an abandon ship drill on August 24, 1944. This tower is 18 feet high. Classes of 50 men went through a continuous drill of climbing the cargo net, descending lines, ascending ladders, and jumping. Note the student with his foot in the face of an instructor in the bottom right.

Gas mask training was conducted for almost all trainees at the station. Sailors had to wear gas masks through a smoke-filled building as part of their final training. Here, final instruction is given on the proper fit and aftereffects of gas on July 5, 1944.

On Saturday, June 5, 1945, a group of 500 German prisoners of war was transported to the station. The first group was transferred from a camp in Aliceville, Alabama. Most of the prisoners were captured in the previous two years and almost all were enlisted personnel. By October 9, 1945, there were 1,645 prisoners of war at NAS Jacksonville. By May 1, 1946, all of them were gone, although many stayed in the United States.

Ens. Jesse Brown was the first African American to be formally designated a naval aviator. He participated in the Naval Aviation Cadet Program and received his Wings of Gold at a graduation ceremony held at NAS Jacksonville on October 21, 1948. During the Korean War, Ensign Brown operated from the USS *Leyte*, CV-32, flying the F4U-4 Corsair fighter in support of the United Nations forces. On December 4, 1950, while on a close air-support mission near the Chosin Reservoir in support of the 7th Marine Regiment, Ensign Brown's plane was hit by enemy fire and crashed. Despite heroic efforts by other aviators, one of whom crash-landed to try and save his squadron mate, Ensign Brown could not be rescued and died in his aircraft. In February 1973, a 3,963-ton Knox class escort ship, the USS *Jesse Brown*, was named in his honor.

This view of the station seawall shows the 100 PBY Catalina seaplanes assigned to the station during World War II. There were four seaplane ramps the large aircraft could use to taxi up on to the parking ramp. Only two of the four seaplane hangars exist today, and only one of the five seaplane ramps has been maintained with access from the St. Johns River to the station parking aprons. The two hangars towards the top of the photograph were demolished in 2008, and Hangar 1122, a new helicopter hangar dedicated on May 5, 2009, took their place. These PBYs all departed the station after the end of World War II, with the last one leaving the station on April 9, 1946.

Three

AIRCRAFT OPERATIONS

A group of new aviation cadets listens intently as a pilot instructor gives final instruction before the men get into their Ryan NR-1 aircraft. An instructor and a student pilot fly in each aircraft. The instructor sits in the backseat, with the student in the front. Ryan built 100 of this model of trainer, all of which were initially flown to NAS Jacksonville.

The first landing on the still unfinished runways took place on September 7, 1940, in this Naval Aircraft Factory N3N-3 biplane. The station's executive officer, Cmdr. V.F. "Jimmy" Grant, was at the controls, with the station's public works officer, Cmdr. Carl Cotter, as his passenger. The neutrality markings on this aircraft are under the wings. The aircraft took off within 15 minutes after landing so that construction workers could get back to work.

The first training flight took place on January 2, 1941. In what was more of a media event than actual training, 25 bright yellow Boeing N2S Stearman aircraft took off with their new student pilots. Someone forgot to get the student pilots their flight goggles, and naval personnel were quickly dispatched to local motorcycle shops with funds to acquire motorcycle goggles before the flight could commence.

Pres. Franklin D. Roosevelt's motorcade drives past a Consolidated P2Y-2 seaplane on the ramp on March 20, 1941. There were a total of nine P2Y-2 and P2Y-3 seaplanes initially assigned to the station, with the first arriving on March 12, 1941. As the first training seaplanes assigned to NAS Jacksonville, they were used to form seaplane-training squadrons VN-14 and VN-15. The last of these aircraft departed on July 10, 1941.

This all-yellow Naval Aircraft Factory N3N-3 is seen in the floatplane configuration on January 7, 1942. The air operations building is in the right background. This two-seat primary trainer biplane was powered by a 235-horsepower Wright R-760-2 Whirlwind 7 radial piston engine. Three of these aircraft were assigned to the station, with one in the seaplane configuration. It was the last biplane in military service when it was retired in 1961.

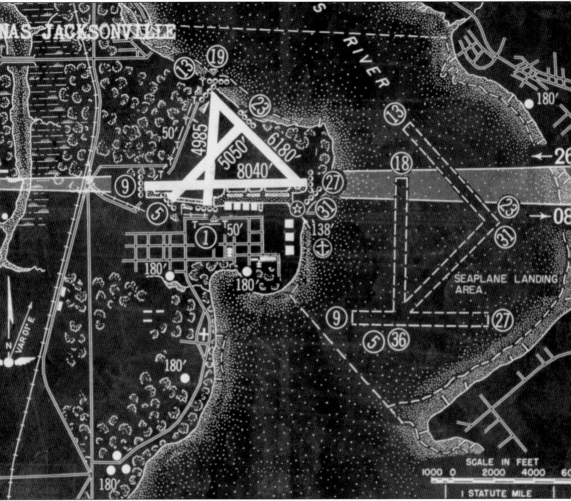

This map shows the runway configurations for the airfield at NAS Jacksonville and the seaplane runway configurations in the St. Johns River in 1941. The seaplane landing and takeoff areas were delineated by marker lights in the St. Johns River. In a *Florida Times-Union* newspaper article in 1942, the base security officer warns the citizens of Jacksonville that continued theft of the seaplane marker lights would result in criminal prosecution. On June 9, 2009, the NAS Jacksonville main runway 09/27 was redesignated 10/28 due to magnetic variation over the past 69 years. This marked the first change since those runways were originally designated. Today, runway 1/19 is only used as a taxiway down to the combat-aircraft loading area, while runway 5/23 is no longer used at all. Runway 13/31 (now 14/32) has a displaced threshold, meaning its length has been considerably shortened.

This remarkable photograph shows aircraft of squadron VN-11B stacked in one of the hangars on September 13, 1941. It was thought that a hurricane was approaching, so maximum use of space was the order of the day. Stearman No. 22 (first row, middle) was the very first plane ever reworked by the Assembly and Repair shops (now the Fleet Readiness Center Southeast). The aircraft with three number designations (left side of photograph) are all Ryan NR-1 trainers, and the planes at the bottom of the photograph are SNJs. All the Stearman and Ryan aircraft were tilted up on their noses to make as much room as possible for aircraft storage during the approaching storm.

This Boeing N2S-3 Stearman biplane was the first aircraft type assigned to the station. On December 24, 1940, a group of 10 new Stearmans landed at the station, coming directly from the factory in St. Louis. On January 2, 1941, Primary Training Squadron VN-11 was established and the first training flight was conducted. There would eventually be 208 Stearman aircraft assigned by December 1941. Thousands of airmen learned primary flight fundamentals by initially flying in this aircraft.

The Ryan NR-1 started arriving at the station in July 1941, and 10 air station pilots flew to San Diego to fly these aircraft back. It was a tedious task, as a gas stop had to be planned every 200 miles along the flight path. They were initially called the Flying Washing Machine by early station pilots, as they barely made it off the ground when taking off. Eventually, all 100 aircraft were sent to the assembly and repairs shops for more powerful engine modifications.

There were 110 Consolidated PBY-1 and PBY-2 model aircraft built for the Navy. NAS Jacksonville initially had 31 of these assigned for patrol bomber training. The first PBY-1 model Catalina aircraft arrived on June 20, 1941, and the PBY-2 models arrived on July 20, 1941. Patrol bomber training was one of the main aviation training programs conducted at NAS Jacksonville throughout World War II, and these aircraft were a familiar sight around the skies of Jacksonville and along the beaches.

In 1943, Consolidated PBY-5 and PBY-5A Catalina aircraft were the dominant PBYs at the station, replacing the earlier models. There were 100 of these aircraft assigned to the station, performing training in gunnery, bombing, and even patrol missions. Here, a PBY-5A is seen being moored on the St. Johns River. At the height of training activities, there were four squadrons of Catalinas based at the station. The last PBY Catalina departed on April 9, 1946.

Above, a Vought OS2U-2 Kingfisher sits on the ramp near a seaplane hangar on January 7, 1942. There were 95 OS2U-2 and OS2U-3 aircraft assigned to the station, which were used for seaplane training. The catapult launch system located on the station pier was also used for launch training. All of the Navy's battleships had a Kingfisher assigned, and training from the catapult for launching off of one of the battleships was conducted at NAS Jacksonville, as seen below. Note the two different paint schemes. The image above shows an all sea blue color and the image below, taken seven months later, shows a two-tone light blue and dark blue combination. The neutrality markings from the earlier photograph were also replaced, since the United States was at war when the photograph below was taken.

A Navy Beechcraft JRB-2, a light transport aircraft, is seen here on the ramp in January 1942. This aircraft was used by senior officers for official transport to other naval installations and for flying to locations for official business.

The first North American Aviation SNJs began arriving at the station in March 1941, and 85 SNJ-3 aircraft were assigned to the station by December 1941. This aircraft was used for intermediate instructional training of pilots. Additionally, an SNJ-3 was modified as the station ready plane. With the introduction of intermediate training, the airfield also had to be enlarged by an additional 2,000 feet in July 1941. This photograph was taken on January 8, 1942.

Although 185 Grumman JRF-5 aircraft were built for the Navy, there was only one aircraft assigned to the station during World War II. It was used as a utility aircraft for official business by the senior aviators and officers at the station, who would take this aircraft for official flights to other air stations and use when traveling to areas where a water landing was necessary.

There were 19 Curtiss SNC-1 Falcon aircraft assigned to NAS Jacksonville when this photograph was taken on January 8, 1942. These aircraft were used for intermediate pilot training. There were 305 Falcons built for the Navy. In late 1942, the NAS Jacksonville–based aircraft were sent to Naval Auxiliary Air Station Lee Field in Green Cove Springs for use in pilot training there.

Corsairs fly in formation (right) and a Vought F4U Corsair catches a wire (below) during a landing at the station in 1944. At the time, the station's new training focus was advanced fighter pilot training. Pilots would learn advanced gunnery, bombing, and fighter tactics at the station. Once they completed that course of instruction, the next stop was usually to a fighter squadron or to an aircraft carrier in the Pacific. Fighter pilot training started on November 18, 1943, with the establishment of Fighter Squadron VF-5. Corsairs would be flown from the station as late as February 1954, when Fighter Squadron VF-44 became the last squadron to switch from the F4U Corsair to the F2H-2 Banshee jet.

The station used this Douglas SBD Dauntless as the station weather plane in 1945. It would go up, scout the flying areas, and report back on weather conditions. Most of the Dauntless dive-bombers were assigned to Naval Auxiliary Air Station Cecil Field, where considerable training took place in dive-bombing techniques in this aircraft.

This Boeing N2S Stearman was the last one attached to the station when this photograph was taken on April 18, 1945. All other Stearman aircraft had long since been transferred to NAS Corpus Christi, Texas. It was kept as a station ready plane, one that could be launched quickly for any task that might be needed. Shortly after the end of World War II, this plane was disposed of, ending the era of biplanes at NAS Jacksonville.

In January 1948, this Howard GH-1 Nightingale was used as a senior officer's utility transport aircraft at the station. The Navy had hundreds of these aircraft built by the Howard Aircraft Corporation during World War II. Other Navy uses were as an instrument trainer or as an ambulance plane. It only remained at the station for about two years.

On November 28, 1944, the first helicopter visited the station. This Sikorsky HNS landed in front of a crowd that had never seen a helicopter and then remained at the station for a few days. The Navy had just acquired four and was evaluating their capabilities with flights to various naval air stations. It would be another five years before a helicopter would be permanently assigned to the station.

The Curtiss SC-1 Seahawk first arrived at NAS Jacksonville on June 11, 1946. Scout observation training was performed in these aircraft. This was the last aircraft in which seaplane training was taught at the base. The Seahawks, of which 577 were produced, were phased out of Naval service starting in 1948, and by 1949, the NAS Jacksonville aircraft were gone.

A Curtiss SC-1 Seahawk seaplane is lifted from the St. Johns River and then placed onto the seaplane ramp at the station on April 18, 1947. Once it had been lifted, personnel attached the wheels to the float, allowing the aircraft to taxi down to either the ramp parking area or into a hangar for needed repairs or maintenance.

These two McDonnell F2H Phantoms arrived at the station on June 10, 1948, drawing huge crowds when they landed. This was the second time a Phantom jet had landed at the station. The first landing was on January 11, 1947, when a Phantom jet made a quick, brief stop to refuel and then went on to set a new speed record on its way to Miami.

The Martin JRM-2 Mars flying boats were the Navy's largest seaplanes at the time. The Navy had five of the flying boats, and two of them, the *Caroline Mars* and the *Philippine Mars*, made frequent trips to NAS Jacksonville, landing in the St. Johns River. Here, the *Caroline Mars* lands, bringing midshipmen to the station for indoctrination training.

On September 26, 1949, when this Lockheed XR60-1 Constitution landed at the station, it was the largest plane to ever land at the station. This plane, bureau number 81563, was the first of only two Constitutions built for the Navy. It was designed as a large transport aircraft. The engines were underpowered for the size of the aircraft, and both were retired from naval service in 1953.

Blimps routinely operated off the coast of Florida during World War II from their base at NAS Glynco, near Brunswick, Georgia. Here, Navy Blimp ZW-2 makes a quick stop at NAS Jacksonville on October 4, 1949, touching down to drop off Lieutenant McCausland. There were two squadrons of blimps operating at NAS Glynco in 1949: the ZP-2 (seen here) and ZP-3.

The second helicopter assigned to NAS Jacksonville was the Sikorsky HUP-1 above. The first helicopter rescue, in a Sikorsky HO3S-1, occurred on July 5, 1950. That helicopter, temporarily assigned to NAS Jacksonville but based at Lakehurst, New Jersey, with helicopter squadron HU-2, was called to rescue a man from a capsized boat east of the station in the St. Johns River. He was stuck in the river mud and rescue boats were having a hard time even getting to him. But when the helicopter arrived, he refused to go into the strange new sight. So, they returned back to the station and left him. The next rescue went a little more smoothly, as Ens. John Neal, assigned to Fighter Squadron VF-74, crashed his F8F Bearcat at Lee Field in Green Cove Springs. He was rescued and brought back to the station. Below, an HUP-2 lands in front of the Air Operation Building on August 25, 1954.

An F2H-2 Banshee is seen here parked on the ramp in front of the Air Operations tower on August 28, 1951. The Banshee was the dominant jet assigned to station-based fighter squadrons at that time. This aircraft was assigned to Fighter Squadron VF-172 and to the aircraft carrier USS *Essex* during the Korean War.

A P2V-2 Neptune assigned to Patrol Squadron VP-18 flies over the St. Johns River with the base in the background on July 3, 1953. The guns can clearly be seen on this model. The Lockheed P2V Neptune was the main antisubmarine warfare aircraft based at NAS Jacksonville from December 1949 until 1967. It was assigned to the newly formed Reserve Patrol Squadron VP-62 in November 1970, which was waiting for its P-3 Orion aircraft.

This Vought F7U Cutlass jet makes an appearance at the base on June 13, 1954. Although no Cutlass aircraft were assigned to squadrons at NAS Jacksonville, they were assigned to NAS Cecil Field. The aircraft, as sleek as it looked, was underpowered and had a bad habit: When the nose wheel hit too hard during an aircraft-carrier landing, it would drive up through the cockpit, usually killing the pilot. They were reworked at the station's Overhaul and Repair Department from September 1955 through 1956.

NAS Jacksonville had one detachment from Utility Squadron VU-10, whose main job was target-towing duties. The plane the men used for this mission was the Douglas JD Intruder. It was painted in bright colors for easy recognition, since other fighters were shooting at the target they towed. The detachment, composed of 25 members, was based at NAS Jacksonville from 1955 through October 1963.

Initial delivery of the Douglas A3D-1 Skywarrior to the fleet took place when the first of five aircraft landed at NAS Jacksonville on March 31, 1956. They were ferried from Naval Air Station Patuxent River, Maryland, and assigned to Heavy Attack Squadron VAH-1, the Smoking Tigers. To the right and behind this A3D (bureau number 135422) are two S2F Trackers and an AF Guardian. This aircraft crashed while assigned to VAH-3, the Sea Dragons, on May 18, 1960.

Attack squadrons have been based at NAS Jacksonville since 1947. They remained until the last attack squadron, VA-176, departed on April 25, 1968. Attack Squadron VA-176 was also the Navy's last piston-engine carrier-based attack squadron and the last to operate the Douglas Skyraider. Here, an A-1H Skyraider of Attack Squadron VA-35 is seen on the ramp on July 3, 1959.

The crew of a Douglas RA-3B Skywarrior aircraft prepares to depart on a mission in the Caribbean on October 1, 1962. Note the middle aviator with a cigar, which were not allowed on the flight line and ramps. Heavy Photographic Squadron VAP-62 flew the Skywarrior while attached to NAS Jacksonville until the squadron's disestablishment on October 15, 1969.

A Sikorsky SH-3A Sea King assigned to the Fleet Angels of Helicopter Combat Support Squadron HC-2 is seen here on the ramp shortly before the squadron was disestablished on September 30, 1977. The squadron arrived at NAS Jacksonville in October 1973 from NAS Lakehurst, New Jersey. During the squadron's 39-year history, it performed a total of 2,318 rescues. The squadron was reestablished in 1987 and is based at Naval Station Norfolk, Virginia, today.

A Beechcraft UC-12B Huron flies over downtown Jacksonville in June 1985. The Huron was a small passenger aircraft used by senior officers at the station. Almost every naval air station had one of these aircraft assigned. Due to funding cuts, the aircraft was removed from service at the station in 2002.

The first Sikorsky SH-60F Seahawk, assigned to Helicopter Squadron HS-3, arrived in September 1991. The Seahawk would replace all of the SH-3 Sea King helicopters at the station by February 2000. Only one squadron, HS-11, still flies this helicopter attached to NAS Jacksonville. This squadron is scheduled to depart in 2014. The SH-60 conducts antisubmarine warfare and has a crew of four.

A Lockheed P-3 Orion enters the aircraft wash area on the flight line ramp. After every flight, the aircraft comes through this wash to remove the corrosive salt spray. The P-3 Orion has been flying in the skies of Jacksonville since 1964 and will continue to be a dominant aircraft until it is finally phased out by the P-8A Poseidon in 2019.

Reserve Squadron VR-62 arrived at NAS Jacksonville on July 25, 2009, from NAS Brunswick, Maine. It flies the Lockheed-Martin C-130T cargo plane. This squadron also provides logistical support to the fleet all over the world. There are four C-130 aircraft assigned to the squadron.

The Boeing C-40A Clipper arrived at the station on August 24, 2002, with a formal dedication ceremony. The new Clippers replaced the four C-9 Skytrains of Reserve Squadron VR-58 within a year. This squadron provides logistical support throughout the world and has been based at NAS Jacksonville since November 1, 1977.

The first Boeing P-8A Poseidon arrived on March 5, 2012, and was introduced with a formal ceremony on March 12, 2012. This new variant of a Boeing 737 jet will replace the current P-3 Orions. Patrol Squadron VP-30, the patrol community training squadron, was the first squadron to receive the new aircraft. Patrol Squadron VP-16 was the first operational squadron to receive the new jets, which replaced all of its P-3 Orions in January 2013.

Four

CRASHES

This Stearman crash is one of 110 aircraft crashes documented in the historical records, which have killed more than 200 personnel. These are crashes that occurred either while landing or taking off from the station, during training flights that originated from or were returning to the station, or involved aircraft from squadrons attached to the station. These records are not all-inclusive, as additional crashes are discovered occasionally in old records or are found when inquiries come into the station about a past crash.

Stearmans were involved in the first crashes at the station in 1941, which caused the first aviator fatality. This was mainly due to the fact they were the dominant aircraft at the station early in 1941 and cadets were learning to fly in them. This Stearman did a soft landing after developing an engine problem about five miles west of the station.

This Stearman crashed fairly violently west of the station. The exact cause of this crash is not known, nor is it described in the aircraft crash files. It is also not known if there were any fatalities in this crash, but the damage to the cockpit area seems pretty severe.

This North American Aviation SNJ-3 (bureau number 01779) developed an engine problem approaching the station and crashed in the woods near the weapon magazines in December 1943. The cockpit is fairly intact even though the aircraft is all but destroyed. Of all the crashes that have occurred at the station, the crash on November 6, 1944, was by far the most tragic. An R4D transport plane, the Navy variant of a DC-3 Skytrain, was ferrying a group of nurses to the station. While on final approach over the St. Johns River, it flew head-on into an F4U Corsair fighter plane that had just departed the station. All personnel on both aircraft, a total of 18 people, were killed. Of note, this crash was not covered by the local Jacksonville papers, as the base's public information officer worked with the press to keep the accident out of the news. Aircraft safety remains a huge priority with the Navy, and tremendous strides have been made to date. Fortunately, aircraft accidents at the station are extremely rare today.

These two Grumman F8F Bearcats were involved in a taxi accident on May 3, 1950, on runway 27. Both aircraft were assigned to Fighter Squadron VF-11, and the tail of one Bearcat was obviously chewed off by the propeller of the other Bearcat, which ran up on it. Plane-taxiing accidents happened quite frequently.

This Vought F4U Corsair was being towed along the perimeter road to the disposal yard when the driver got the wing too close to the fence, which clipped the wing and did considerable damage to the aircraft. Fortunately, it was being disposed of, so the only damage requiring repair was done to the fence.

This North American Aviation SNJ-5 had a landing-gear failure upon landing on June 13, 1947. Here, the crash crane, nicknamed "Tilly," lifts the aircraft so the landing gear assembly can be inspected and the aircraft can be removed from the runway.

A Vought F4U-5 Corsair sits on the flight line after landing and experiencing an engine fire, which was extinguished using foam by the airfield crash fire truck, on March 4, 1950. Engine fires were not uncommon in the early 1950s. Fortunately, this pilot escaped unhurt and the damage to the aircraft was minor.

A McDonnell F2H-1 Banshee's port landing gear failed to go down as the jet returned to the station on July 19, 1950. The pilot was forced to do a belly landing, resulting in minor damage to the plane. This jet was assigned to Fighter Squadron VF-11, home-based at the station at that time.

This dump truck turned a corner too sharply along a line of Lockheed TV-1 Navy jets parked on the flight line on July 7, 1954. A jet wing tank pod went through the windshield of the dump truck when the driver cut the corner too close, causing the truck to swerve to the left, where it hit another TV-1 jet before coming to a stop.

This Lockheed NC-121 Super Constellation, assigned to the Hurricane Hunters Squadron VW-4, went off the end of the runway on October 11, 1957. Air bags were used to lift the wing and landing gear, which sank into the soft dirt. Then rocks were used to fill in depressions underneath to make a harder surface so the plane could be towed back onto the runway.

On Saturday, July 16, 1966, a Fairchild C-119 transport took off from the station and suffered an engine fire almost immediately. At 7,000 feet, all 34 personnel donned their parachutes and jumped, some for the first time. Three minutes later, the empty plane crashed near Otis Road and the Atlantic Coast Line Railroad tracks in Jacksonville. It took rescue teams many hours to find all of the men, many of whom were dangling in trees by their parachutes, but no lives were lost in the crash. (Associated Press Wire photograph.)

A Lockheed P2V Neptune assigned to the Hurricane Hunters Squadron VW-4 at NAS Jacksonville is seen here partially submerged on December 11, 1956. While towing the aircraft from its parked position on the seawall parking ramp, the tow bar broke, allowing the plane to roll backward, eventually going into the St. Johns River.

On January 10, 1978, a McDonnell-Douglas AV-8A Harrier jet assigned to Marine Corps Squadron VMA-542 was at the station preparing for landing. As the plane made the transition from flight to hover mode for the landing, it suddenly veered off and crashed. The pilot safely ejected out of the sideways aircraft, but the plane was totally destroyed. (Photograph by R.E. Kling.)

Three Convair C-131 Samaritan transport planes are seen here on the station ramp in 1981. These aircraft would routinely arrive at the station transporting personnel and equipment. The closest plane in the photograph was the Guantanamo Bay station's transport plane at the time. Scheduled contracted flights to and from Guantanamo Bay still fly on Tuesdays and Saturdays from the station.

On April 30, 1983, the Naval Station Guantanamo Bay Convair C-131 transport plane left NAS Jacksonville heading to Cuba. It developed an engine fire shortly after takeoff. The pilot immediately tried to return to the station. Parts of the burning engine hit cars on Old King's Road in the Mandarin area of Jacksonville. The left wing finally separated, and the aircraft crashed into the St. Johns River just short of the approach runway, killing 14 personnel. There was one survivor.

This Sikorsky SH-3 Sea King, assigned to station-based Helicopter Squadron HS-1, is hoisted out of the St. Johns River and placed onto the fuel pier at the station in 1989. The helicopter had a tail rotator failure and was able to land in the St. Johns River, deploy its floats, and get towed back to the station by rescue boats.

This Lockheed P2V-5 Neptune did a partial wheel-up landing at NAS Jacksonville when it arrived from Kissimmee, Florida, on March 5, 1992. The port main landing gear failed when the plane was ready to land. Remarkably, the pilot landed on two wheels, then turned the plane into the grass just as he set it down, only doing very minor damage to the aircraft. The plane, restored by Patrol Squadron VP-5 in 2011, can be seen today in the station's static display park. (Photograph by R.E. Kling.)

Five

THE BLUE ANGELS

The Blue Angels Flight Demonstration Team was formed at NAS Jacksonville on April 24, 1946. The members of the initial team are, from left to right, Lt. Al Taddeo (solo pilot), Lt. (jg.) Gales Stouse (flying Beetle Bomb and spare pilot), Lt. Cmdr. Roy M. "Butch" Voris (flight leader), Lt. "Wick" Wickendoll (right wing), and Lt. Mel Casidy (left wing).

The first aircraft assigned to the Blue Angels were these F6F Hellcats. Note there are no dots after the U and the S in front of the name Navy on the sides of the aircraft. The letters were made of gold leaf, a very expensive process at a time when the military was winding down after World War II. The aircraft are seen here at Craig Field in Jacksonville just prior to their first public demonstration on June 15, 1946. (Photograph by Mernard Norton.)

The second aircraft assigned to the Blue Angels were Grumman F8F Bearcats. This photograph was taken at Craig Field in Jacksonville during the 1947 air show. The Bearcats were much more maneuverable than the Hellcats and were well liked by the early Blue Angel pilots. Both Butch Voris and Al Taddeo of the original Blue Angels team have come back to visit the station for recent air shows. (Photograph by Mernard Norton.)

An SNJ was initially assigned to the Blue Angels to simulate a Japanese fighter. On August 26, 1946, it was replaced by this F8F Bearcat, named Beetle Bomb. Its job was to initially attack the Blue Angels aircraft, which would split into two formations to engage. Beetle Bomb would simulate being hit, with smoke and a dummy thrown out of the back of the plane to simulate a "kill." This part of the demonstration ended in 1949 and Beetle Bomb faded into history when it crashed while taking off at a show in Pensacola in 1950.

The Blue Angels team walks away from its aircraft after a practice at NAS Jacksonville on June 17, 1947. The men are, from left to right, Lt. Robert Thilen, Lt. Charles Knight, Lt. Robert Clark, Lt. Cmdr. R.E. "Dusty" Rhodes (flight leader), and Lt. (jg.) W.C. May. The Blue Angels were based at NAS Jacksonville until November 8, 1948, when they were relocated to NAS Corpus Christi, Texas.

The Blue Angels fly by the station in their F9F-5 Panthers during an air show on January 9, 1954. The Panther jet was first flown by the team in 1949 and continued to fly off and on through the 1954 season. An updated Cougar jet was used starting in 1955.

Miss Jacksonville contestants greet the Blue Angels pilots before their performance at NAS Jacksonville's 23rd Open House Anniversary on October 11, 1963. The Blue Angels were flying the F11-F-1 Tiger jets when this photograph was taken.

Six

A&R, O&R, NARF, NADEP, NAVAIR, AND FRCSE

The Assembly and Repair Department was first established in offices located in Hangar 101 on March 14, 1941. The main entrance is seen here. The hangar is one of 11 structures at NAS Jacksonville designed by Albert Khan, a famous architect of the day. Most of those 11 buildings are still in use today. He designed all of the original hangars, the brig, a torpedo workshop, the barracks, and the mess hall and galley.

Civilians and military personnel of the Assembly and Repair Department sand propellers in the propeller shop on September 23, 1942. Thousands of civilian and military personnel repaired aircraft assigned to the station during World War II.

Artisans stitch a horizontal stabilizer from a Boeing N2S Stearman in this early photograph. Both women were assigned to the fabric/wing shop at the Assembly and Repair Department in 1941. Most of the station aircraft at that time had fabric wings.

The Assembly and Repair Department women's basketball team held the title of station champions when this photograph was taken on July 6, 1944. There were hundreds of WAVES working at various shops during World War II. They performed many functions related to aircraft repair and maintenance.

In this early photograph of the Assembly and Repair shops, three Martin PBM Mariners are seen in various stages of rework on September 21, 1944. These aircraft flew in conjunction with the Consolidated PBY Catalinas, conducting submarine patrols off the coast of Florida. They were visitors at NAS Jacksonville.

The floor layout of Hangar 101 at the Assembly and Repair Department is sectioned off by aircraft type in this September 15, 1944, photograph. In the center are Lockheed PV-1 Ventura patrol bombers, and to the left are Douglas SBD Dauntless dive-bombers.

A hangar full of Vought F4U Corsairs in various stages of rework is seen here on February 6, 1947. A dominant Navy and Marine Corps fighter during World War II, Corsairs were reworked at the station shops from the first induction in 1943 until the last one was completed in March 1954.

Assembly and Repair Department technicians hoist and install an engine on a Vought F4U Corsair on August 23, 1945. The worker on the stand (left) is John S. Robinson, the grandfather of coauthor Ronald M. Williamson. This photograph was discovered while going through old editions of the base newspaper, the *Jax Air News*.

The Relations Department at the station controlled civilian payroll. The payroll truck is seen here in front of the main hangar at the Overhaul and Repair Department at the station on June 27, 1947. Personnel are lined up to get their payroll, which was paid in cash.

The first jet inducted to the Overhaul and Repair Department was this McDonnell FH-1 Phantom jet, which arrived in July 1949. The personnel in the photograph are unidentified. Since that time, the shops at this facility have reworked almost every type of jet the Navy has flown.

The Assembly and Repair Department changed its name to the Overhaul and Repair Department on July 22, 1948. Helicopter repairs started in 1951. By the time helicopter repairs were transferred to the Naval Air Rework Facility at NAS Pensacola in late 1967, more than 20 different models of helicopters had been repaired. A total of 2,123 aircraft went through the facility, with the last helicopter completed on December 7, 1967.

A small Bell TH-13 Sioux helicopter is seen here in the Overhaul and Repair shops in 1966. Helicopter repairs in the 1950s and 1960s ranged from this helicopter, the smallest, to the huge Sikorsky CH-37 Mojave, capable of carrying 35 armed Marines into combat.

The Overhaul and Repair Department helicopter-blade testing structure is seen to the right of center in this 1958 photograph. This structure tested thousands of blades, spinning them at full power from 1952 until it was finally removed in 1967.

The Overhaul and Repair Department civilian flight-test crew checks the blades' "pitch of rotor" on a Sikorsky HRS-1 helicopter on September 26, 1952. In the test, the pole is slowly moved towards the spinning blades; if one blade is out of pitch, it will hit the section along the top side of the pole and leave a chalk mark on that blade. If that occurs, the blades are stopped and the pitch can be adjusted. The same basic procedures are employed today, except that a laser tracker is used.

Artisans from the Overhaul and Repair Department prepare an R4D-5 Skytrain for Adm. Richard Byrd's first naval Operation Deep Freeze expedition to the Antarctic on May 30, 1955. One problem that needed to be overcome was the engine oil cooler positions. Situated at the bottom of the plane, they were subject to damage from flying snow and ice during takeoff and landing. It was finally decided to just remove them, as they were not necessary in the cold climate.

On February 14, 1966, the Overhaul and Repair Department was designated as a rework site for the Douglas A-4 Skyhawk and the Vought A-7 Corsair. The first A-4E Skyhawk is seen here awaiting induction just a few months later.

This interior shot of Hangar 101 shows the A-4 Skyhawk repair line. Numerous Douglas A-4 Skyhawks are seen here in various stages of rework. More than 1,200 of these aircraft were reworked, from the first induction in April 1966 until the final aircraft was completed in August 1973.

A Pratt & Whitney J-52 engine goes though a final inspection prior to being placed in an A-4 Skyhawk in this 1967 photograph. The J-52 was developed in the mid-1950s and is still used to power the Northrop Grumman EA-6B Prowler aircraft, which are reworked at the same hangar today.

The Overhaul and Repair Department was renamed the Naval Air Rework Facility on April 1, 1967. Additionally, this activity became a separate tenant command on the station and was no longer a station department under the commanding officer. This name lasted until March 31, 1987, when the name changed to the Naval Aviation Depot.

A fire truck gets into position on April 8, 1975, to fight the fire that eventually consumed the east cooling tower for the Naval Air Rework Facility's new jet engine test cell. The tower, which was under construction, had to be rebuilt.

The first Vought A-7 Corsair inducted for rework at the Naval Air Rework Facility arrived on September 21, 1967. It was assigned to Attack Squadron VA-174. The Corsair was one of the longest-lasting aircraft ever worked at the facility. The last one was inducted in September 1990. Foreign military sales saw the A-7 return to the facility to be reworked for a few more years.

The Naval Air Rework Facility had a family day on October 29, 1970. Here, an unidentified father and his two sons look into the cockpit of a Douglas A-4 Skyhawk in Hangar 101. Family days have been held over the years to give family members an opportunity to see what really goes on in the facility. Normally, visitor access is restricted.

On May 26, 1981, a Northrop Grumman EA-6B Prowler aircraft crashed on the flight deck of the aircraft carrier USS *Nimitz*. The crash and subsequent explosion killed 14 aircrewmen, injured 45 more, and damaged or destroyed 11 aircraft. Six of the damaged aircraft were brought to the Naval Air Rework Facility for repairs, including this Vought A7-E Corsair.

The first McDonnell Douglas F/A-18 Hornet (pictured) was brought into the Naval Air Rework Facility in August 1984 for evaluation. Even today, F/A-18 Hornets can be seen at the NAS Jacksonville Fleet Readiness Center Southeast site and at an outlying site located at the former NAS Cecil Field.

The seaplane observation and control structure, located on the roof of the Naval Air Rework Facility's Hangar 140 since 1942, is seen here being removed in conjunction with hangar renovations in the early 1980s.

A Lockheed S-3 Viking waits for a tow truck after final modifications had been made in late 1986. The Naval Air Rework Facility artisans did repairs on this aircraft from 1986 to 1990, and then again on three special aircraft from 2009 to March 2010.

Artisans from the Naval Air Rework Facility stand in front of the first scheduled depot level maintenance (SDLM) F/A-18 Hornet on December 8, 1987. Hornets have been continually reworked since this first one, with the latest workload being center barrel replacement. The sign reads, "Put the beer on ice 'cause this one's done."

Seven

DECADES OF GROWTH AND CHANGE

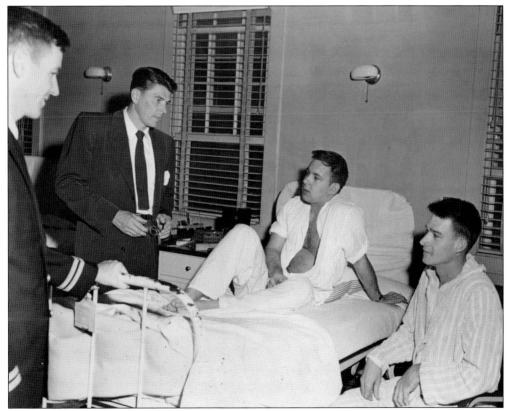

Movie star and future president Ronald Reagan is seen here visiting patients at Naval Hospital Jacksonville on January 20, 1951. Many stars of the day would visit the facility and talk with the military patients.

With the transfer of this book on February 16, 1953, Patrol Squadron VP-741 was disestablished and Patrol Squadron VP-16 was established at the station. The Eagles of VP-16 remain at the station today, even though several other patrol squadrons have been disestablished in the past 15 years.

Cmdr. G.H. Ghiresguire (right), the commanding officer of Patrol Squadron VP-3, admires a model of a P2V Neptune built by Petty Officer Burns, also of VP-3, on June 1, 1953. In a base newspaper clip just a few weeks later, the *Jax Air News* asked if anyone had any information on the missing model. It is not known if it was ever recovered.

In one of the most colorful ceremonies ever held at the station, six Landing Support Ships Large (LSSLs) were transferred from the US Navy to the Italian Navy on July 25, 1951. These transfers were in accordance with the Mutual Defense Assistance Program. The program, started by Pres. Harry Truman in 1949, was meant to bolster the capabilities of United States allies during the start of the Cold War.

Sailors load guns on a Lockheed P2V-5 Neptune patrol aircraft assigned to Patrol Squadron VP-5 on July 23, 1951. The guns used during this period were the Navy's M3 cannon, which fired 20-millimeter cartridges. In the mid-1950s, the Mk12 system replaced the M3. This system had a lighter projectile with a bigger charge, for better muzzle velocity and a higher rate of fire.

NAS Jacksonville departments, as well as tenant commands located at the station, would build elaborate floats for parades in downtown Jacksonville, starting with the station's fifth anniversary in 1945 and continuing through the early 1960s. The float seen here is from a 1950s parade. The two floats on the opposite page participated in the station's fifth anniversary parade in 1945. The station no longer provides any floats for Jacksonville parades.

This float, designed by Naval Auxiliary Air Station (NAAS) Cecil Field personnel, reflects the mission of that station during World War II. Dive-bomber training was the main focus of the training conducted by personnel. NAAS Cecil Field was under the commanding officer of NAS Jacksonville during World War II and through 1952. It was finally designated a separate naval air station on July 1, 1952.

Like Cecil Field, Mayport was also designated as a naval auxiliary air station during World War II, also under the direction of the commanding officer of NAS Jacksonville. NAAS Mayport was used as an emergency divert field for aircraft returning to NAS Jacksonville after doing either training or carrier-landing practice off the coast of Jacksonville. In addition to being an emergency airfield, NAAS Mayport was also a location for crash boats.

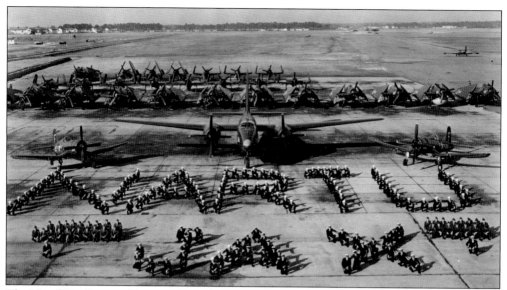

The Naval Air Reserve Training Unit shows its aircraft and personnel in this 1954 photograph. The Naval Air Reserve Program started on July 1, 1946, at Naval Auxiliary Air Station Cecil Field, in Hangar 14. With so many experienced Navy pilots returning to civilian life after the war, this gave many an opportunity to stay involved with naval aviation. The Reserve Training Unit moved from Cecil Field to NAS Jacksonville in August 1946 and held an open house that attracted 6,000 visitors.

Aircraft of the Naval Air Reserve Training Unit are seen here in front of Hangar 113 in 1962. The aircraft are, clockwise from top, an S2F-1 Tracker, a T-34B Mentor, a P2V-5F Neptune, an A4D-2N Skyhawk, an HSS-1N Seabat, an SNB-5 Navigator, an R5D-3 Skymaster, and a TV-2 (T-33B) T-Bird.

116

The Naval Air Technical Training Center commenced operations at NAS Jacksonville in September 1942. Before that, it was only known as the Trade Schools. The headquarters building, seen here, was located on Yorktown Avenue between Biscayne and Child Streets. The Naval Air Technical Training Center was closed on February 21, 1974, and this building was demolished shortly thereafter. The site is now the home of Patriots Grove Park, a tribute to Navy personnel who have won the Medal of Honor since the start of World War II. Lee Harvey Oswald attended the schools at the station in 1961.

During the station's 20th anniversary celebration on October 14, 1960, the airfield was dedicated as John Towers Field. Admiral Towers's widow, Pierrette Anne Towers (below), was at the ceremony. The plaque that was unveiled is in front of Building 118, the air operations building. Rear Admiral Towers was naval aviator number three and was instrumental in almost every aspect of early naval aviation. He was present during the station's commissioning on October 15, 1940, and visited often. Rear Admiral Towers developed the coveted Wings of Gold that all naval aviators receive upon successful completion of flight training. He planned and led the first Atlantic air crossing in May 1919 and then commanded the Navy's first aircraft carrier, the USS *Langley*, from January 1927 to August 1928. He was promoted to the rank of vice admiral and commanded the Naval Air Force US Pacific Fleet in World War II.

On Sunday, July 11, 1971, an air show had just ended at the station and traffic was exiting as a severe thunderstorm raged. Suddenly, the Naval Air Technical Training Center galley was on fire. The building (above) burnt to the ground and half a block of structures were destroyed, with damages estimated at $1 million. It was finally determined that an electrical short, not lightning, as had first been suspected, was the cause of the fire.

A former prisoner of war, Lt. Cmdr. John S. McCain III arrives at NAS Jacksonville on March 6, 1973, following his release from captivity. McCain, a future US senator from Arizona and presidential candidate, was attached to a squadron from NAS Cecil Field prior to going to sea. He returned to NAS Cecil Field after the Vietnam War and became the executive officer and commanding officer of Attack Squadron VA-174, the Hellrazors, in the mid-1970s.

Lt. Fred Krift and Lt. Judith Ann Neuffer reflect on the events that followed a flight as they stand alongside an Orion WP-3A weather reconnaissance aircraft attached to Weather Reconnaissance Squadron 4 (VW-4). Neuffer was one of the first female pilots in the Navy.

Commander, Tactical Air Atlantic was established at NAS Jacksonville in a ceremony on May 1, 1973. On July 12, 1974, it was moved to Naval Station Norfolk, Virginia. Commander, Fleet Air Jacksonville was also disestablished, and, in place of those two commands, Commander, Sea Based Antisubmarine Warfare Wings Atlantic was established. The command is now called Commander, Navy Region Southeast.

Asst. Secretary of State Sidney Sober (second from right) greets King Hussein of Jordan upon his arrival at the station on May 4, 1975. Also on the trip was the shah of Iran. They were escorted to NAS Cecil Field, where an impressive display of military fighter aircraft was available for viewing.

The station had a 53-year history with trains. In one of the last photographs of trains at the station, in 1994, the station's switcher engine is seen here in front of the static display area on view on Yorktown Avenue. This engine is now in the Florida Gulf Coast Railroad Museum in Tampa.

The senior leadership of NAS Jacksonville went to the Pentagon in 1991 to receive the Commander-in-Chief's Annual Award for Installation Excellence. This award recognizes the best base in the Navy. The command won the award a second time in 2011, and again in 2012. Capt. Kevin Delaney (first row, center), the commanding officer of the base, is posing for this photograph with station personnel on May 24, 1991.

Patriots Grove Park, consisting of 81 monuments dedicated to the Navy personnel who have received the Medal of Honor since the start of World War II, was dedicated on April 19, 1996. Seen here at the podium is Capt. Robert Whitmire, base commanding officer. The three Medal of Honor recipients present at the dedication are, from left to right, Lt. (jg.) Thomas J. Hudner, HM2 (Hospital Corpsman) Donald E. Ballard, and BM1 (Boatswain's Mate) James E. Williams. Also on the stage, but not pictured, was Congressman Charles E. Bennett, who gave the keynote address.

On this wet day, October 24, 1997, the commanding officer's plane from Sea Control Squadron VS-24 arrived direct from a deployment aboard the USS *John F. Kennedy*. Cmdr. Roy Ivie's landing began the transition for six S-3 Viking squadrons to move from the closing NAS Cecil Field to NAS Jacksonville. Other squadrons that arrived included VS-30, VS-24, VS-22, VS-32, and VQ-6. By January 29, 2009, all six squadrons were disestablished.

On June 26, 2009, this waterspout formed in the St. Johns River just south of the base. The station had recently installed a new Big Voice system for important announcements, and base commanding officer Capt. Jack Scorby put out the following message: "There is a waterspout near the base. All personnel are requested to shelter in place!" Of course, the opposite happened, as some base personnel grabbed their cell phones and cameras and went to the river to take pictures of the waterspout, which passed harmlessly by.

The Patrol Squadron VP-8 Fighting Tigers are seen (above) receiving the traditional fire department water salute as they formally arrive as the second squadron transferred from NAS Brunswick, Maine, on May 27, 2009. The first squadron that arrived, a week earlier, was Special Projects Patrol Squadron VPU-1, known as the Old Buzzards (below). VPU-1 was also the first squadron located in the just-completed Hangar 511, built for the transferring squadrons. VPU-1 arrived with no advanced announcement or fanfare. VPU-1 had a disestablishment ceremony on April 27, 2012, and then combined with VPU-2, located at the Marine Corps Air Station at Kaneohe Bay, Hawaii. The men and women of VPU-1 flew some of the most sophisticated aircraft and were relied upon to fly the most dangerous and challenging missions in the maritime patrol and reconnaissance community. Following VP-8 and also transferring from NAS Brunswick were the VR-62 Nomads, on July 25, 2009; Patrol Squadron VP-10 Red Lancers, on December 5, 2009; and Patrol Squadron VP-26 Tridents, on June 3, 2010.

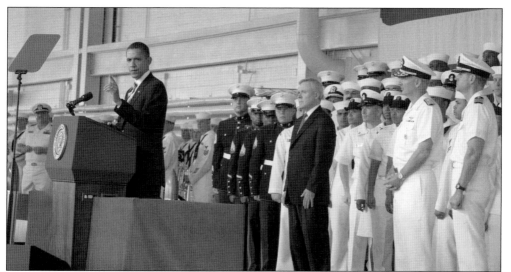

Pres. Barack Obama speaks in front of a full hangar at the station on October 6, 2009. Base commanding officer Capt. Jack Scorby is in the first row on the far right. The new helicopter hangar used for the speech, Hangar 1122, was perfect for this event. It was still under final review and had not been formally given to the government yet. Thousands of base personnel attended to see their commander-in-chief in person.

Shortly before President Obama's speech in Hangar 511, Helicopter Squadron Maritime HSM-70 was the first squadron to move in, flying the new Sikorsky MH-60R Seahawk. There will eventually be five HSM squadrons, with 51 aircraft, assigned to the station. This helicopter provides a wide variety of antisubmarine warfare as well as surface rolls and can be operated from a wide variety of ships. The former HS squadrons at NAS Jacksonville, and the HSL squadrons at NS Mayport, will convert to HSM squadrons as they receive the new helicopters.

INDEX